Create Your Magic Squares

Constructions – Oddities – Teaching Ideas

Yann Le Bail

Contents

Introduction..4
  A few definitions...5
Part 1 : The great ancestor: the Luò Shū (洛書)..................................7
  Introduction to the Luò Shū..8
    Teaching ideas..8
  The Luò Shū and 1089...10
    Teaching ideas..12
  The Luò Shū, probabilities, and intransitivity............................12
    Teaching ideas..14
Part 2: As big as you want (but odd)...17
  The Siamese method...18
    First variation..18
    Second variation..25
    Teaching ideas..32
Part 3: We tackle even orders..33
  Simple method for order 4: we swap..34
    Teaching ideas..36
  General Method..36
    Teaching ideas..42
  Roger Bowley's magic square...43
    Teaching ideas..46
  The knight's tour..49
Part 4: Where we end the even orders...51
  Al-Kharaqī's method..52
    Teaching ideas..58
Part 5 : How about we multiply?...59
  Some operations...60
    Teaching ideas..64
Part 6: The Magic of panmagic squares of order 5............................65
  Magical patterns galore..66
  Constructions...69
    First variation..69
    Second variation..72
  A tour de force...74
    Magic constant multiple of 5..74
    Any magic constant..76

  Teaching ideas...77
 Yann Le Bail's magic squares...78
  The method..78
  Teaching ideas...79
Conclusion...81
Bibliography..82

Introduction

From the first officially known in China over 2000 years ago, related to numerology, to mathematical studies conducted by computers, magic squares have always fascinated the mind. Numerological tools, lucky charms, talismans, spellcasters, or simple objects of study, we find historical traces of them almost everywhere in the world, from ancient China to pre-Columbian civilizations, via Europe.

This book aims to share my humble experience in this field through an almost playful approach, with a pedagogical perspective.

This book does not presuppose any mathematical knowledge other than the notion of whole numbers, addition, multiplication, etc., so it is accessible to an audience from middle school, but, I hope, it will delight adults of all ages. The few calculations required can be done by hand by an elementary school student, which makes the construction of the magic squares in this book accessible to everyone.

Far from being exhaustive, this book is more of an introduction to magic squares and the intrinsic beauty of these objects at the crossroads of numbers and geometry.

Our journey will start with the smallest squares, moving to squares with odd-sized sides, considered the easiest to construct without calculation. For squares with even-sized sides, again, apart from a single simple calculation that can be done by hand, nothing else will be needed to create your square, not even a calculator, except perhaps colouring pencils.

Stemming from history or more modern discoveries, each theme covered is punctuated by remarkable, curious or sympathetic magic squares, whether for the professional or amateur mathematician. We will also find that they can be used by a teacher, a parent, or simply by who wants to go further in the subject and related subjects.

A few definitions

A magic square is a square array containing numbers. But that's not all. For it to be magic, the sums of the numbers of its rows, columns and diagonals must all be equal. For example a square completely filled with numbers 2 is a magic square: the addition of the rows, columns and the two diagonals each time giving the result 6. But let's face it: this magic square is not very interesting.

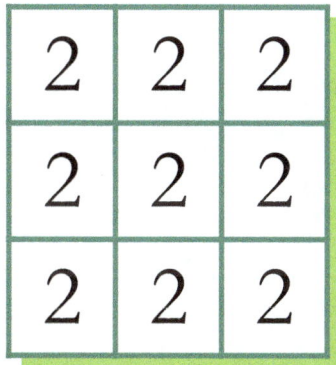

Generally, the magic square will be given additional constraints, such as the numbers contained in the square having to be all different, or even worse, all sequential. These constraints are not mandatory however, and so the square above is indeed a magic square, a so-called "trivial" magic square.

The number 6 obtained by performing these sums is called the *magic constant*. Similarly, the number of rows or columns is called the *order* of the square, so here we have an order 3 magic square.

IIt is quite easy to calculate the magic constant of a magic square: just add the contents of all its cells, and divide by its order. So, if you want to construct a magic square of order 4 with all the numbers from 1 to 16, the sum of these numbers being 136, the magic constant will be 136 ÷ 4, that is, 34.

6

All we have to do is bring a pencil and squared paper and we are now ready for our little mathematical journey.

Part 1:
The great ancestor: the Luò Shū (洛書)

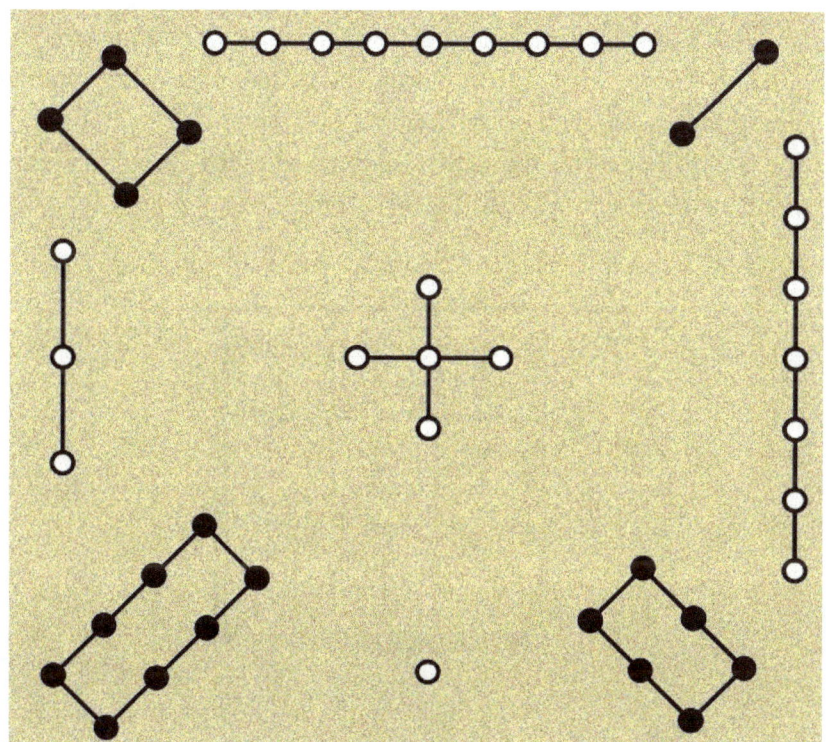

The Luò Shū.

Introduction to the Luò Shū

Known since the second century BC in China, the Luò Shū was originally an object of numerology, very unscientific. But in the 12th century AD, it was studied by the mathematician Yang Hui (楊輝) (c. 1238 - 1298), as a mathematical object stripped of its esotericism. Before Yang Hui, Arab mathematicians had been studying magic squares since the 7th century.

This magic square is the only one of order 3 containing all the numbers from 1 to 9, if we ignore the different possible symmetries. Does this mean that other than its size and history, it offers little interest to the curious mind? In reality, its simplicity hides few treasures that may surprise more than one.

4	9	2
3	5	7
8	1	6

Teaching ideas

The first obvious pedagogical application is to present the magic square with an empty grid and ask the students to fill it in as it is, without hints, after having presented the properties of magic squares. This will make it possible to use the vocabulary of sum, row, column and diagonal, to introduce the notion of magic constant, among others. Since the exercise can be difficult for younger students, we can give some clues:

- give the magic constant to be obtained (15), which sum of rows, columns or diagonals;
- give the number in the central cell (5);
- even indicate that a diagonal is 4, 5, 6.

We can also make the students find a way to calculate the magic constant, by discovering the general method of adding all the numbers in the magic square, and dividing by the number of rows or columns. Here, the sum of the numbers from 1 to 9 divided by 3.

Thus, each step can be given resolved to the students or given to them to discover.

Then, in the whole class, we can collect the different solutions and compare them in order to talk about the notions of symmetries and rotations and show that, up to a symmetry, these solutions are in reality only one.

Finally, it is possible to start a by showing a picture of the Luò Shū and let the students guess what it represents.

The Luò Shū and 1089

But we are not done with this magic square yet. Indeed, we can have fun multiplying all the numbers in the Luò Shū by 1089 in order to discover a very nice curiosity.

4356	9801	2178
3267	5445	7623
8712	1089	6534

Indeed, the square obtained, even if we lose the sequentiality of the numbers it contains, ends up with very singular properties.

Note that the fact that we obtain a magic square is not surprising: everything being multiplied by 1089, the square obtained is proportional to the original square, the magic constant is then $15 \times 1089 = 16{,}335$.

On the other hand, the attentive reader will have noticed that the thousand digits alone (in black) again form a Luò Shū[1]. Similarly,

---

[1] The reader will forgive me for simplifying the expression "the numbers represented by each of the thousands digits". I will allow myself this type of shortcut several times to lighten the text.

the unit digits (in green) form a Luò Shū, but this time symmetrical with respect to the central cell.

What about the other numbers? The answer is that we obtain with the digits of the hundreds (in red) then with the tens (in blue) two magic squares equivalent, to a symmetry, to the Luò Shū subtracted by 1 in each cell, their magic constant being 12.

Thus, all these properties have the consequence that any combination of two or three digits, as long as they represent the same rank in the starting numbers, i.e. the same colour, will form a new magic square, for example with black and blue:

45	90	27
36	54	72
81	18	63

Note that here the numbers obtained are all multiples of 9. This is no coincidence, but the explanation is left to the reader to discover.

Finally, we can also shuffle the numbers, provided we do it the same way in each cell. It is not useful to take all four digits of each number, two or three will do. So, taking the order black, green, red, and not using blue, we get the magic square:

463	918	281
372	554	736
827	190	645

If these simple rules are followed, we will always get a magic square. Amazing, right? How many different magic constants can we get with all these different ways to shuffle and all these possible choices? The answer is left to the reader.

Teaching ideas

We can make the students think about the consequences on the magic constant of multiplying an entire magic square by the same number (like 1089 above), or even of adding or removing the same number to all the squares of the magic square (seen higher by subtracting 1, for example). Thus the properties of distributivity, proportionality or linearity can be introduced. They will be useful to us in the last part.

The Luò Shū, probabilities, and intransitivity

After this title designed to frighten non-mathematicians, it is appropriate to reassure. The notions and curiosities that we are going to discover here are very simple.

Let's go back to our Luò Shū, but this time we'll colour the columns:

4	9	2
3	5	7
8	1	6

Besides being pretty (and patriotic for the French), these colours highlight a very particular property that we are going to examine with the help of a game.

Imagine that I have three blue cards in my hand, each having the numbers 4, 3 and 8 respectively, and that the player has the white cards 9, 5 and 1. If we play one of our cards at random, who will be more likely to win (winning being having a higher value card)?

There are 9 possible encounters. But the reader's 9 will beat my three cards, his 5 will beat my 4 and my 3. So that gives me 5 possible defeats out of the 9 probable encounters. In short, clearly I have no interest in playing this game, where the odds are against me in the long term.

Unless, for the second round, I choose the white cards. What will the reader do then? Take the weaker blue cards? Obviously not, and he will be right, so he will take the red cards. And if my 9 beats his three cards well, the only other possible win I have left is my 5 against his 2. In short, I end up with 4 wins against 5 probable losses, and the player will still win.

Sore loser that I am, for the last round, and to save the little honour I have left, I then choose the red cards. What will the reader do? He will obviously not take the white cards, whose weakness we now know against the reds. But what about the blue ones? In a final display of cunning wisdom, the player then chooses the blue cards. Indeed, he saw that his 8 will beat my three cards, that his 4 and his 3 will beat my 2. In short, 5 wins for the reader out of 9 possibilities.

But how? Blue isn't the weakest colour? Red isn't the strongest? What is this sorcery?

We are in fact in a situation known as *intransitivity*. The one found in the game of rock-paper-scissors.

"What if instead of columns we consider rows, does that work too?" the reader will wonder. For reasons of symmetry, the principle is the same and will work, except that we have to colour our cards following the horizontal stripes:

4	9	2
3	5	7
8	1	6

The result is obviously less patriotic for a Frenchman but will please the inhabitants of Schleswig-Holstein.

Teaching ideas

In addition to teaching the correct pronunciation of Schleswig-Holstein, [ˌʃleːsvɪç ˈhɔlʃtaɪ̯n], we can calculate the probabilities of victories or defeats for each column or row over another, introduce the notion of transitivity (if a < b < c then a < c) and hence the intransitivity seen here.

We can also have win tables drawn up as follows to find out which row or column is the strongest:

	9	5	1
4	White	White	Blue
3	White	White	Blue
8	White	Blue	Blue

Blue column vs. white column win table: white column wins 5 out of 9 times

It is also possible in a sampling and simulation activity, to work with dice instead of working with cards, by taking two sides of a six-sided die for each number. We can even have the students design these dice themselves. Working with dice will also make it possible to present non-classical dice such as George Sicherman's two dice or James Grime's five intransitive dice.

As a gift to the reader, here are the column situation dice patterns:

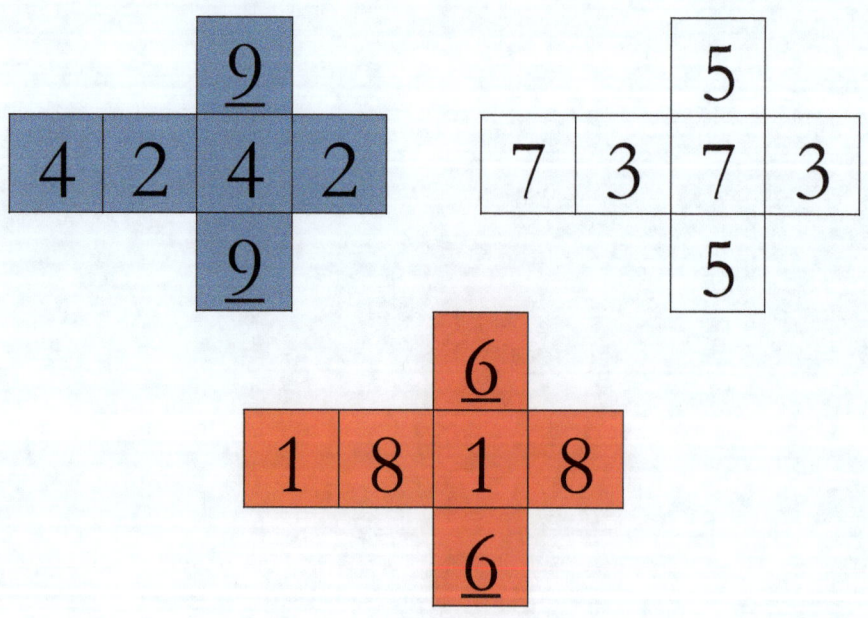

And the ones of the row situation:

Part 2:
As big as you want (but odd)

17	24	1	8	15
23	5	7	14	16
4	6	13	20	22
10	12	19	21	3
11	18	25	2	9

Ce quarré est essentiellement different de celuy d'Agrippa : la Methode de Bachet ne s'y accommode pas aisement ; & au contraire la

Excerpt from Du Royaume de Siam : Tome 2 *by Simon de La Loubère.*

The Siamese method

In volume 2 of his book <u>Du Royaume de Siam par Monsieur de La Loubère, envoyé extraordinaire du Roy auprès du Roy de Siam en 1687 et 1688</u>, Simon de La Loubère indicates a general method for creating magic squares of odd order (of side 3, 5, 7, etc.) of any size, counting by 1 from 1. He actually presents two methods, both known as the Siamese method or the La Loubère method.

Yes, we are talking about any size, as long as that size is an odd number. And this without any calculation! If the reader has the urgent desire to create a magic square of order 2025, he will be able to do so in 5 minutes (that is to say, he will be able to start, but finishing this square of 4,100,625 cells is another story).

One concept that will be needed and reused throughout these pages is that of *going around*. Indeed, when we move according to certain rules in a magic square, it is possible to "go beyond the edge" and leave the square. In this case, we will imagine that we are teleporting from right to left, from top to bottom and vice versa. But all this will be clearer with examples.

First variation

The rules for the first variation of the Siamese method are as follows:

— if necessary, *go around*;
— start counting on the centre cell of the top row;
— move counting one cell diagonally up to the right (⤴);
— if this diagonal move is impossible, move down one cell (⬇).

Let's take a square of order 5 as an example and place the 1 as indicated in the rules:

We want to go to the top right; going to the right is fine, but we're at the edge at the top, so we'll have to *go around*, and then we end up at the bottom:

Then we continue to the top right and easily place the 3:

For the 4, we'll go beyond the right edge and end up on the left column as we *go around*:

Then we place the 5 with no problem:

		1		
	5			
4				
				3
			2	

For the 6, we cannot go to the top right, the cell being occupied by the 1. The rules specify that in this case, we go down one cell:

		1		
	5			
4	*6*			
				3
			2	

We are now able to continue diagonally, which we do without issue until the 8, and, *going around* for the 9 and then for the 10:

		1	*8*	
	5	*7*		
4	6			
10				3
			2	*9*

Unable to place the 11 diagonally, we put it under the 10:

		1	8	
	5	7		
4	6			
10				3
11			2	9

Then we finish the diagonal:

		1	8	*15*
	5	7	*14*	
4	6	*13*		
10	*12*			3
11			2	9

The location of the 16 diagonally is taken, *going around*, by the 11, so we write the 16 below the 15:

		1	8	15
	5	7	14	*16*
4	6	13		
10	12			3
11			2	9

Then we continue, *going around*, until the 20. Note here the position taken by the 17:

17		1	8	15
	5	7	14	16
4	6	13	*20*	
10	12	*19*		3
11	*18*		2	9

Finally, by inscribing the 21 under the 20 and finishing diagonally while *going around* when necessary, we complete this magic square of order 5 and magic constant 65:

17	*24*	1	8	15
23	5	7	14	16
4	6	13	20	*22*
10	12	19	*21*	3
11	18	*25*	2	9

Second variation

The rules for the second variation of the Siamese method are as follows:

- If necessary, *go around*;
- Start counting on the cell just below the centre cell of the square;
- Move counting one cell diagonally down to the right (⇘);
- If this diagonal move is impossible, move two cells down (⇓⇓).

Let's take a square of order 5 as an example and place the 1 as shown in the rules:

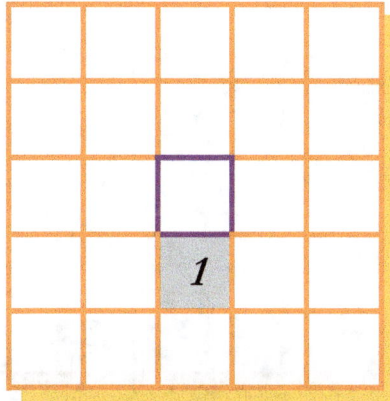

We want to go down to the right, and place the 2:

		1		
			2	

Then we *go around* and easily place the 3:

				3
		1		
			2	

For 4, we go beyond the right edge and end up, *going around*, on the left column:

Then we place the 5 with no problem:

For the 6, we cannot go to the lower right. The rules specify that in this case, we move down two cells:

				3
4				
	5			
		1		
	6		2	

We are now able to continue to 10 diagonally by *going around* for the 7 and the 10:

		7		3
4			*8*	
	5			*9*
10		1		
	6		2	

Unable to place the 11 diagonally, we put it two cells below the 10, *going around*:

11		7		3
4			8	
	5			9
10		1		
	6		2	

Then we finish the diagonal:

11		7		3
4	*12*		8	
	5	*13*		9
10		1	*14*	
	6		2	*15*

The location of the 16 diagonally is taken, *going around*, by the 11, so we write the 16 two cells below the 15, *going around*, of course:

11		7		3
4	12		8	*16*
	5	13		9
10		1	14	
	6		2	15

Then we continue, *going around*, until the 20:

11		7	*20*	3
4	12		8	16
17	5	13		9
10	*18*	1	14	
	6	*19*	2	15

Finally, by inscribing the 21 two cells below the 20 and finishing diagonally, while paying attention to the position of the 23, we finally complete this magic square of order 5 and magic constant 65:

11	*24*	7	20	3
4	12	*25*	8	16
17	5	13	*21*	9
10	18	1	14	*22*
23	6	19	2	15

So in one method we actually have two methods that create two different magic squares for a given odd order:

17	24	1	8	15
23	5	7	14	16
4	6	13	20	22
10	12	19	21	3
11	18	25	2	9

11	24	7	20	3
4	12	25	8	16
17	5	13	21	9
10	18	1	14	22
23	6	19	2	15

Teaching ideas

The creation of magic squares by this method is a very good application of the notion of program and algorithm. The notion of *going around* can also introduce the basics of modular arithmetic, or lead to talk about the remainder of Euclidean division.

Students can be asked if this method also works for order 3. Does it generate a Luò Shū? How to explain that both variants get two symmetrical Luò Shū? We can try to lead them to discover that, in a 3 × 3 square, going up or down one space is the equivalent of going down or up two spaces when *going around*.

Part 3:
We tackle even orders

Albrecht Dürer's magic square, detail from Melencolia I.

Simple method for order 4: we swap

While the methods for even orders may seem complicated in general, for order 4 a few swaps are enough:

We fill a square with the numbers from 1 to 16:

1	2	3	4
5	6	7	8
9	10	11	12
13	14	15	16

Without moving the corners, we swap the right and left edges:

1	2	3	4
8	6	7	*5*
12	10	11	*9*
13	14	15	16

Still without moving the corners, we swap the top and bottom edges:

1	*14*	*15*	4
8	6	7	5
12	10	11	9
13	*2*	*3*	16

Finally, we swap in column then in row, or directly diagonally, the numbers of the central square:

1	14	15	4
8	*11*	*10*	5
12	*7*	*6*	9
13	2	3	16

And so we get a magic square of order 4, with magic constant 34. Moreover, the sum of the four numbers of the central square is 34, as well as the sum of the four corners, among others.

Teaching ideas

It is possible to have the students look for other magic patterns (the central square and the four corners are some) with a sum of 34.

Students could also be asked to talk about Albrecht Dürer's work, *Melencolia I* (see illustration at the beginning of this part), the symbols that this work contains, and to discuss the magic patterns of his magic square, as well as a method to construct it.

Finally, by noticing that the writing of numerals today is different from that of Dürer's time, students can be asked to research the history and evolution of Indo-Arabic numerals.

General Method

The previous method of swapping rows and columns seems to no longer work for orders greater than 4. Luckily for us, Arab mathematicians devised non-computational methods to create magic squares of order as large as one wants, but multiples of 4. Among these is a method from an anonymous 12th century manuscript that uses only colours, except for one detail: we will need an additional piece of information, namely to know by what number we multiplied 4 to obtain the order of our magic square. If our magic square is of order 8, the number sought is *2*, because $4 \times 2 = 8$. In short, nothing insurmountable.

Let's start directly with a simple example: the order 4. So we multiplied 4 by... *1* to get 4. So this *1* will be our working base.

We draw a 4 × 4 square:

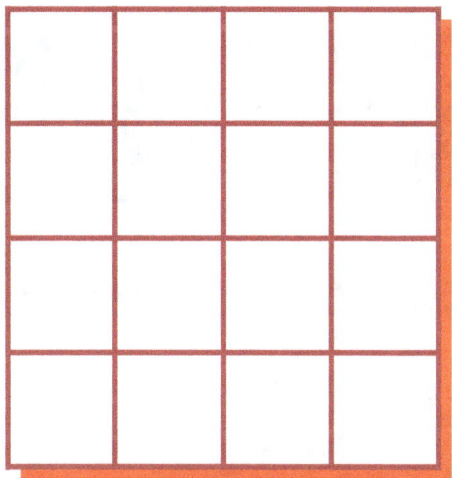

We'll work in the upper left quadrant (for example):

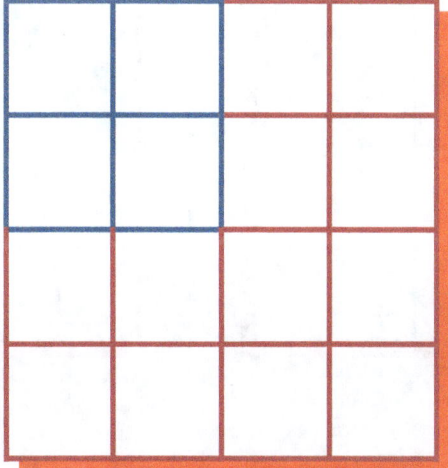

In this quadrant we are going to colour cells in such a way as to obtain *1* coloured cell per row and *1* coloured cell per column exactly. Why *1*? Because *1* is our working base, let's not forget.

Here, there are only two ways to get the desired result:

Now all we have to do is perform a vertical axis symmetry on the colours:

Then a horizontal axis symmetry:

Then we just have to count from right to left then from top to bottom, writing only on the coloured cells. Be careful, we will still count the white cells even if nothing is written there:

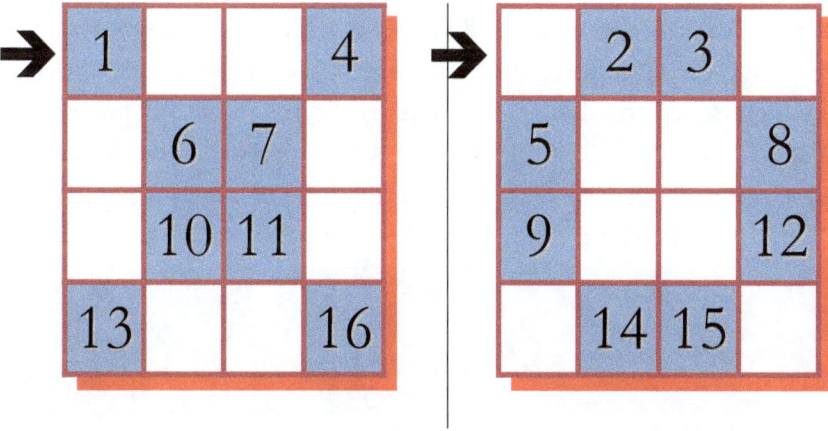

Finally, and this is the last step, we just have to count in the other direction starting from the lower right corner, from right to left and from bottom to top, this time writing only in the white cells:

1	15	14	4
12	6	7	9
8	10	11	5
13	3	2	16

16	2	3	13
5	11	10	8
9	7	6	12
4	14	15	1

And here are two beautiful magic squares of order 4.

Let's go crazy: let's try another example: 8 × 8. Here, our working base is the number *2*, because 4 × *2* = 8, as seen above. We draw our magic square, then, we fill in the cells of the upper left quadrant so as to obtain *2* coloured cells per row and per column. Diagonals don't matter. It is at this stage that we are most likely to make mistakes (especially if our working base is large). A sure-fire way to get it right is with a chequerboard pattern. But it's less pretty…

So we'll go for this staircase pattern:

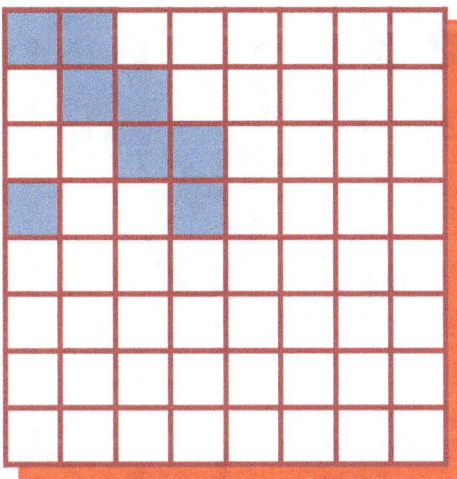

Then we apply the symmetries:

Then we fill in the numbers, first in one direction:

1	2					7	8
	10	11			14	15	
		19	20	21	22		
25			28	29			32
33			36	37			40
		43	44	45	46		
	50	51			54	55	
57	58					63	64

Then the other one:

1	2	62	61	60	59	7	8
56	10	11	53	52	14	15	49
48	47	19	20	21	22	42	41
25	39	38	28	29	35	34	32
33	31	30	36	37	27	26	40
24	23	43	44	45	46	18	17
16	50	51	13	12	54	55	9
57	58	6	5	4	3	63	64

And there you go!

Teaching ideas

Apart from the fact that filling in the magic squares is a good exercise in counting and then calculation (we must check that we have not made a mistake) inherent in all the constructions of magic squares in this book, and without forgetting the calculation in advance of the magic constant, we can also lead the students to reflect on the notions of multiple and divisors, symmetry and even broaden the subject on the writings systems of the world in general, and in which direction they are carried out (Latin alphabet, Arabic alphabet, traditional Chinese, etc.).

We can also introduce panmagic squares, also called pandiagonal magic squares, among others, where the broken diagonals, put together by *going around*, also have the magic constant as their sum. For example:

6	9	7	12	6	9	7	12
3	16	2	13	3	16	2	13
10	5	11	8	10	5	11	8
15	4	14	1	15	4	14	1

These squares have the particularity, among other things, that we can take the last column and place it first, the bottom column and place it at the top, etc. to always get a panmagic square. Students can be asked to explain why.

Roger Bowley's magic square

Here is a very interesting magic square in that it highlights new properties due to the way numbers are written. This time, we will write the numbers as if we had only four digits: 1, 2, 5 and 8. We then write all possible two-digit numbers. What luck! There are exactly 16 of them:

11 ; 12 ; 15 ; 18 ; 21 ; 22 ; 25 ; 28 ; 51 ; 52 ; 55 ; 58 ; 81 ; 82 ; 85 ; 88.

Can we use these numbers to get a magic square? The answer is yes, and here is Roger Bowley's magic square:

11	22	58	85
55	88	12	21
82	51	25	18
28	15	81	52

With a magic constant of 176. Note also that the sum of the four central cells is also the magic constant.

What more can be said? Precisely, the modern era has brought us what the ancients did not know: the seven-segment display of numbers, that of old calculators and digital watches and clocks.

So if we rewrite Roger Bowley's magic square this way, the whole specificity of that magic square comes out:

11	22	58	85
55	88	12	21
82	51	25	18
28	15	81	52

Because the particularity of this writing is that some of its digits are symmetrical by central symmetry, in particular the digits 1; 2; 5 et 8 and therefore remain unchanged by turning them upside down.

So, turning upside down this magic square, we get another one with the same characteristics:

25	18	5	82
81	52	15	28
12	21	88	55
58	85	22	11

Note that if the digits are unchanged by this half-turn, this is not necessarily the case for the numbers they write. Thus, 25 becomes 52 by turning upside down. It was therefore not guaranteed to obtain a magic square in this way.

But that's not all, because these numbers have the particularity of remaining numbers after a vertical axis symmetry, the two squares above then remain magic squares even in a mirror (or through the sheet of paper where they are written), becoming:

28	82	55	11
15	51	88	22
81	25	12	58
52	18	21	85

58	12	81	25
85	21	52	18
22	88	15	51
11	55	28	82

They too have the same magic constant of 176 and the central square equal to this constant. A magic square in every way!

Unfortunately, it is not possible to create panmagic squares of order 4 having the symmetry properties of Roger Bowley's square.

Teaching ideas

A very playful activity is to work with 16 playing cards, of four values and four different suits, for example all the jacks, queens, kings and aces, of a normal deck, and to have the students recreate the magic square by Roger Bowley. It is indeed interesting to develop the vocabulary of playing cards, especially for probabilities exercises that they might encounter later.

- We start with a puzzle: place the cards on the table, in a 4 × 4 square, so that each row, each column and each of the two diagonals contains only one of each value and each colour. The word "sudoku" may help students understand the concept. After a few minutes, the students begin to solve the puzzle. It is possible to do this activity without playing cards, but it is then necessary to make two tables: one for the cards placed, which at the end will contain the solution sought, and one for the cards used, to avoid using the same card twice.

A possible solution:

Note: Students can be helped to find a solution by giving them the following construction method: fill in the first column, then the two opposite corners (there are only two cards that can go in these corners once the column is completed, each offering a different valid solution), then work logically to fill in the rest.

– Once a solution has been found, the students are offered to write it in a table but by numerically coding each card as follows:

Aces become:	10	Spades become:	1
Jacks become:	20	Hearts become:	2
Queens become:	50	Clubs become:	5
Kings become:	80	Diamonds become:	8

Thus, the queen of hearts becomes 50 + 2 = 52.

The resulting square is, for example:

11	22	58	85
55	88	12	21
82	51	25	18
28	15	81	52

This square is either Roger Bowley's square (like here) or another one with exactly the same properties!
– It is then enough to have the square written in seven-segment characters to make the amazed students discover all its symmetries.
– We can try to make them guess the total number of different squares that can be obtained with this playing cards method. The method of construction indicated in the note above can also serve as a clue.

We can also talk about base 4, which is the idea behind this method and behind Roger Bowley's square, and thus introduce the notion of number bases.

The knight's tour

Before closing this part, I will answer the question that the reader might ask about the problem of the knight's tour, a question studied in particular by the mathematician Leonhard Euler.

In the classic knight's tour problem, we are on an 8 × 8 square chessboard and the goal is to visit all the squares in this chessboard by moving like a knight in chess, without ever visiting the same square twice. By numbering the sequence of squares visited by the knight, we obtain this kind of board:

42	59	2	17	40	15	22	63
3	18	41	60	21	64	39	14
58	43	20	1	16	23	62	37
19	4	57	44	61	38	13	24
56	45	6	29	12	25	36	51
5	30	55	48	33	52	11	26
46	7	32	53	28	9	50	35
31	54	47	8	49	34	27	10

Can we then, among the 19,591,828,170,979,904 possible tours, find a description of a magic square?

The answer is unfortunately no. We can find semi-magic squares, where the sums of the rows and columns are indeed the magic constant of 260, but not the diagonals. There are 140 of these semi-magic squares that can be toured by a knight, counting all symmetries, but no magic square. The square above is an example of this. Pity…

Part 4:
Where we end the even orders

3D representation of a magic square of order 6 by Yann Le Bail.

Al-Kharaqī's method

At the end of the 11th century, the Persian mathematician Abū Muhammad 'Abd al-Jabbār al-Kharaqī (Persian: بهاءالدین ابوبکر محمد بن احمد بن ابوبشر) (? –1138), presents a method with colours to create magic squares of even orders, but not multiples of 4. That's great, that's what we're missing.

As for the method for orders multiples of 4, we need to know the number which will be our working base. This number is easy to find. Just subtract 2 from the order and then divide by 4. For 14, we therefore find *3* as a working base: (14 − 2) ÷ 4 = *3*.

Another way to look at it is that 14 is a multiple of 4 plus 2 (14 = 12 + 2), and we are looking for the working base of this 12 as in the previous part: 12 = 4 × *3*.

Finally, by dividing 14 by 4, we find *3.*5, by ignoring the decimal part, we also find our working base *3*.

Let's see this method for order 10. Our working base is then… *2*, because (10 − 2) ÷ 4 = *2*. The first step is to work in the upper left quadrant:

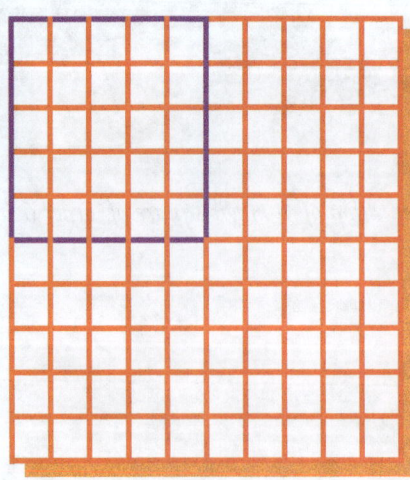

On the top row of this quadrant, we choose a cell that we put in one colour (red, for example), and another cell in another colour (blue). Please note: the leftmost cell is forbidden (×). On the other hand, the other cells can be chosen freely, for example, in a patriotic spirit of the Frenchman that I am:

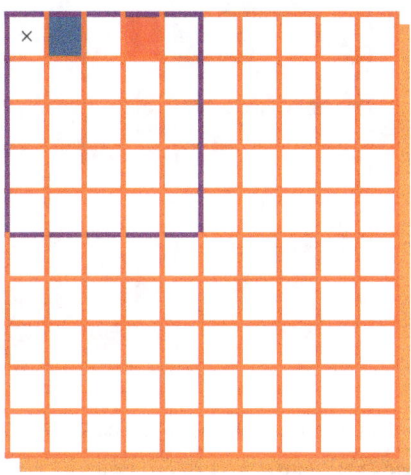

Then, in the remaining cells of this row, we colour *2* cells with a third colour (green). Here we are talking about our working base *2* (unlike the previous step). This time, the leftmost cell is not forbidden (which is fortunate, because in the case of order 6 we wouldn't have enough space...):

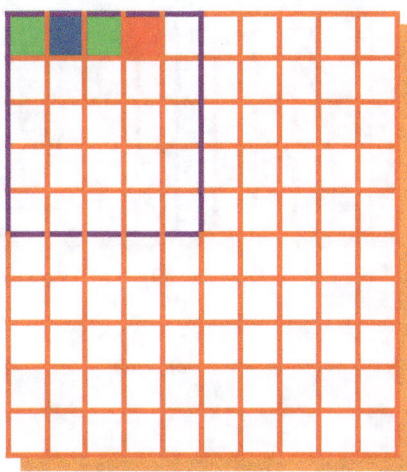

Then, we extend these colours diagonally down to the right, without leaving the quadrant and *going around* when necessary:

Now all that remains is to perform some symmetries. The first of the first colour (red) with a vertical axis:

The second of the second colour (blue) with a horizontal axis:

Finally, we will deal with the third colour, which, the reader will have noticed, has a different status from the other two. It is first subjected to a vertical axis symmetry:

Then horizontal:

Phew! The hardest part is done! Indeed, we only have to write the numbers. We'll start with red, our first colour. Since the red has been mirrored to the *top right*, we start at the *bottom left*, horizontally, counting all the cells, left to right then bottom to top, but writing only in the red cells. Yes, I know, there is nothing to write from 1 to 50. We get this:

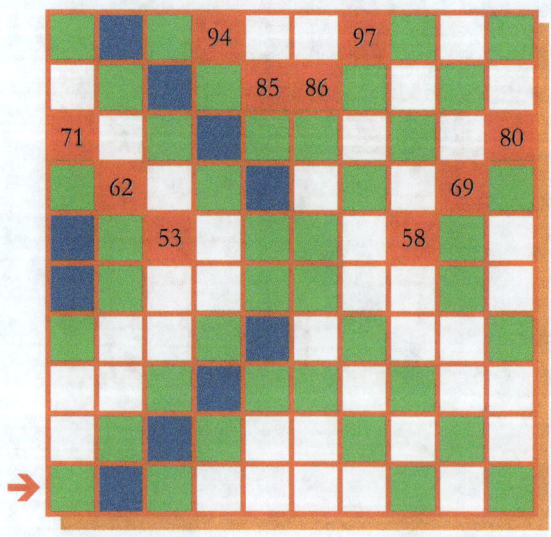

We start again with the blue cells, which have been mirrored to the *bottom left*, so we start counting from the *top right*, moving right to left, then up and down, this time only filling in the blue cells:

For our special colour, green, we won't do anything very special: we start at the top left and go left to right and top to bottom, only filling in the green cells, of course:

Finally, the remaining cells are completed from bottom right, right to left, then bottom to top:

1	9	3	94	96	95	97	8	92	10
90	12	18	14	85	86	17	83	19	81
71	79	23	27	25	26	74	28	72	80
31	62	68	34	36	65	37	63	69	40
50	42	53	57	45	46	54	58	49	51
60	52	48	47	55	56	44	43	59	41
61	39	38	64	66	35	67	33	32	70
30	29	73	77	75	76	24	78	22	21
20	82	88	84	16	15	87	13	89	11
91	99	93	7	6	5	4	98	2	100

And we're done! We now know how to create magic squares of any size, and this, apart from finding our working base, without any calculations or equations.

Teaching ideas

The first teaching idea is once again inspired by the search for the working base, which can lead to exercises on multiples, divisors, and the remainder of Euclidean division.

Another work can be done on the history of mathematics and the Islamic Golden Age.

Part 5 :
How about we multiply?

Magic square of order 2 with sequential numbers 1 to 4.
No, the image is not missing: no such magic square exists.

Some operations

We saw in the first part that we could multiply a magic square by a number, which we did with 1089, and thus obtain a new magic square. However, this one loses the sequentiality of its numbers. The magic constant is also multiplied by the said number.

Even more simply, we can add the same number to all the boxes of a magic square to obtain a new one. This time, the sequentiality is preserved by this operation. We will leave it to the reader to determine what then becomes the magic constant.

Even more, we can add two magic squares of the same size. Here again, the sequentiality is no longer guaranteed. The magic constant, for its part, is the sum of the two initial magic constants.

Can we imagine another operation on magic squares that allows us to obtain new ones? The answer is yes. There is a form of multiplication. Its interest is that the magic square obtained is of order the product of the two initial squares. Better: the sequentiality is preserved.

Here's how: Let's take two magic squares, for example the Luò Shū and Dürer's magic square:

4	9	2
3	5	7
8	1	6

16	3	2	13
5	10	11	8
9	6	7	12
4	15	14	1

To multiply them, we have to proceed in several steps.

First step: choose one of the two. This choice is not trivial, as we will see later.

Let's choose the Luò Shū.

4	9	2
3	5	7
8	1	6

Step Two: Subtract 1 from all cells.

3	8	1
2	4	6
7	0	5

Step three: The second square (Dürer's) has 16 squares. So we multiply everything by 16.

48	128	16
32	64	96
112	0	80

Step four: We transform each cell into 16 identical ones.

48	48	48	48	128	128	128	128	16	16	16	16
48	48	48	48	128	128	128	128	16	16	16	16
48	48	48	48	128	128	128	128	16	16	16	16
48	48	48	48	128	128	128	128	16	16	16	16
32	32	32	32	64	64	64	64	96	96	96	96
32	32	32	32	64	64	64	64	96	96	96	96
32	32	32	32	64	64	64	64	96	96	96	96
32	32	32	32	64	64	64	64	96	96	96	96
112	112	112	112	0	0	0	0	80	80	80	80
112	112	112	112	0	0	0	0	80	80	80	80
112	112	112	112	0	0	0	0	80	80	80	80
112	112	112	112	0	0	0	0	80	80	80	80

Fifth and final step: in each group of 16 identical boxes, we will add Dürer's square.

64	51	50	61	144	131	130	141	32	19	18	29
53	58	59	56	133	138	139	136	21	26	27	24
57	54	55	60	137	134	135	140	25	22	23	28
52	63	62	49	132	143	142	129	20	31	30	17
48	35	34	45	80	67	66	77	112	99	98	109
37	42	43	40	69	74	75	72	101	106	107	104
41	38	39	44	73	70	71	76	105	102	103	108
36	47	46	33	68	79	78	65	100	111	110	97
128	115	114	125	16	3	2	13	96	83	82	93
117	122	123	120	5	10	11	8	85	90	91	88
121	118	119	124	9	6	7	12	89	86	87	92
116	127	126	113	4	15	14	1	84	95	94	81

And there you have it. From squares of orders 3 and 4, we obtained a square of order 12.

This method therefore allows to create very large magic squares, but cannot generate all possible orders. Thus, it cannot generate a square of order 8 with all its sequential numbers, because, as indicated at the beginning of this part, there is no non-trivial magic square of order 2. As soon as the decomposition of the order of the square requires a 2, this method will be found wanting. It therefore does not allow order 8 (= 2 × 4), nor 10 (= 2 × 5), nor 14 (= 2 × 7), etc.

On the other hand, when it works, it allows us to create two different squares. Indeed, in the first step, we chose the Luò Shū. If we had chosen Dürer's magic square, the multiplication would then have generated another very different square. Clearly, this method of multiplying magic squares is not *commutative*.

Teaching ideas

Generating magic squares by this method is quite laborious, but can be interesting as a group work for motivated students. One can work on two Luò Shū, to obtain a smaller square of order 9.

By comparing the two squares obtained, the one above and the one by choosing Dürer's square in the first step, we can make students recall or discover the notion of commutativity of operations in mathematics.

Finally, like most of the methods in this book, this one can be a very good programming exercise for students who are quite advanced in this area.

Part 6:
The Magic of panmagic squares of order 5

Artistic representation of a Le Bail's magic square with mise en abyme. *Of questionable aesthetic taste, it has the advantage of hiding its centre from the reader in order to keep the suspense until the end. By Yann Le Bail.*

We have already tackled the question of panmagic squares of order 4. They do not exist for all orders, but for order 5, we can very easily construct panmagic squares having very, very nice properties that the we can even turn it into a magic trick or a little speed contest between students. But let's not anticipate too much.

Magical patterns galore

Here is a magic square of order 5:

22	5	8	11	19
13	16	24	2	10
4	7	15	18	21
20	23	1	9	12
6	14	17	25	3

It is easy to verify that it is magic, with a magic constant of 65. However, it was not constructed using the Siamese method. We will see its construction method in a moment. What is remarkable about this square is that it is panmagic: its broken diagonals also have a sum of 65, *going around*, for example the red or blue broken diagonals:

22	5	8	11	19
13	16	24	2	10
4	7	15	18	21
20	23	1	9	12
6	14	17	25	3

22	5	8	11	19
13	16	24	2	10
4	7	15	18	21
20	23	1	9	12
6	14	17	25	3

But where this magic square shows its full potential is when we add the centre and the four corners together: again we find 65!

22	5	8	11	19
13	16	24	2	10
4	7	15	18	21
20	23	1	9	12
6	14	17	25	3

And that's far from everything, since this same centre and the middle cells of the four sides: 65 again!

22	5	8	11	19
13	16	24	2	10
4	7	15	18	21
20	23	1	9	12
6	14	17	25	3

Even better: Let's pick a random number from the square. Let's say... 18. Then let's add the four numbers around it to form a "+" sign:

22	5	8	11	19
13	16	24	2	10
4	7	15	18	21
20	23	1	9	12
6	14	17	25	3

We find again 65, and this, whatever the number chosen in the square, by means of *going around* if we are too close to the edge.

The reader wants more?

Never mind: instead of the "+" sign, let's draw the "×" sign, it still works!

22	5	8	11	19
13	16	24	2	10
4	7	15	18	21
20	23	1	9	12
6	14	17	25	3

There are other magical patterns in this square, but the ones listed here are the most notable. They will serve us to do a tour de force (or magic trick) in a moment.

But for now, let's learn how to construct it, or other squares of the same type.

Constructions

First variation

As with the Siamese method, here we have two variations. Be careful though, these only work to create panmagic squares of odd order when they exist. Otherwise, we won't even get a simple magic square. We can easily verify that they will not lead anywhere for order 3, for example.

The method is close to the Siamese method:

— If necessary, *go around*;

— Start counting from any cell;

— Move counting one cell to the right, and two cells up (⇨⇧⇧), like the move of a knight in chess;

— If this move is impossible, move one space diagonally up to the left (↖).

Let's see what we get. For example, we start here, a probable sequel to the second version of the Siamese method:

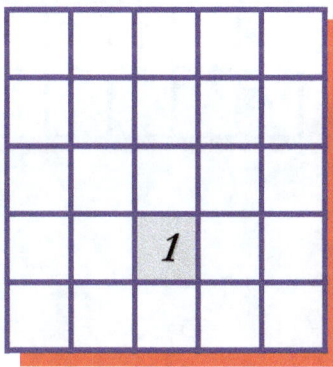

Then we continue with our chess knight move, not forgetting to *go around* when necessary:

	5			
			2	
4				
		1		
				3

After writing the 5, the knight move would take us back to the location of the 1, so we do the other move, diagonally, and *going around*, lay our 6:

	5			
			2	
4				
		1		
6				3

We then continue like a knight until 10:

	5	*8*		
			2	*10*
4	*7*			
		1	*9*	
6				3

Then diagonally again to place the 11:

	5	8	*11*	
			2	10
4	7			
		1	9	
6				3

And we continue like this until the end, going diagonally for 16 and 21:

22	5	8	11	*19*
13	*16*	*24*	2	10
4	7	*15*	*18*	*21*
20	*23*	1	9	*12*
6	*14*	*17*	*25*	3

And there you have it, our panmagic square of order 5!

Second variation

This second variant is even closer to the Siamese method, but it is easier to discover when done in front of an informed audience. We will see that this is not ideal when we do our tour de force. In the meantime, here are the rules:

- If necessary, *go around*;

- Start counting from any cell;

- Move counting one cell to the right, and two cells up (⇨⇧⇧), like the move of a knight in chess;

- If this move is impossible, move down one cell (⇩).

Let's see what we get. We start the same way as before, so we can compare with the other variation:

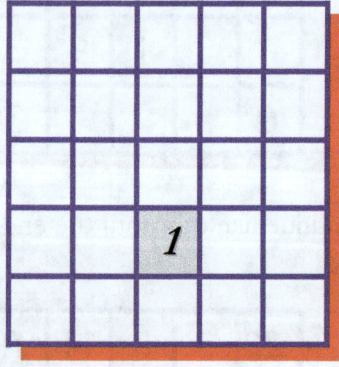

The beginning is the same:

	5			
			2	
4				
		1		
				3

But everything changes from the 6:

	5			
	6		2	
4				
		1		
				3

And so on:

	5			*9*
	6		2	
4			*8*	
10		1		
11		*7*		3

Until the end:

17	5	13	21	9
23	6	19	2	15
4	12	25	8	16
10	18	1	14	22
11	24	7	20	3

A tour de force

At this point of the book, the reader may wonder what is the point of learning two variations of a method to construct a panmagic square of order 5, when it is perfectly reasonable to imagine that we can simply learn one of these magic squares by heart. This question is justified, especially since being able to start where we want in the square does not motivate us to learn any of these methods. However…

Imagine that we could, mentally and with just a sheet of paper, produce such a square, with its cute little patterns of "+", "×" and all, but with *as a magic constant a number chosen by a third party*. And this, I reassure the reader, without great effort of mental calculation.

Magic constant multiple of 5

The principle is simple. If we start at 1, the magic constant is 65, we know that. And if we start at 0? Just try to see that we have deducted 1 from each number in the grid. In other words, we deducted 5 from each row, each column, etc. The magic constant then becomes $65 - 5 = 60$.

Let's say we want to create a square of magic constant 75. So how do we do that? Warning: we are going to do (a few) calculations.

If we start at 0, the magic constant is 60, and each time we increase the starting number by 1, the magic constant increases by 5. For 75, how much has it increased? 75 = 60 + 15. Our magic constant has increased by 15. This represents an increase in the starting number of 15 ÷ 5 = 3. So we'll start our magic square at 3:

24	7	10	13	21
15	18	26	4	12
6	9	17	20	23
22	25	*3*	11	14
8	16	19	27	5

And voilà!

What about numbers below 60? The principle is exactly the same, but we will start with a negative number. For example, if 50 is the requested number, it means that we have reduced our magic constant from 60 to 50, i.e. a decrease of 10, i.e. an increase of − 10. Which again means that our starting number will be negative equal to − 10 ÷ 5 = − 2:

19	2	5	8	16
10	13	21	− 1	7
1	4	12	15	18
17	20	*− 2*	6	9
3	11	14	22	0

Anyway, if you don't like negative numbers, just ask the third party for a "big enough" number, or some other shameless trick to get a number above 60.

Any magic constant

What if the magic constant requested by this third party is not a multiple of 5? Don't panic, the method is simpler than before. We fill in the magic square for the magic square with the magic constant multiple of 5 just below, then we adjust with the last 5 cells. For example, if 78 is requested, we start the magic square for 75, but without filling the last 5 cells, the next cell to be filled being the red one:

	7	10	13	21
15	18		4	12
6	9	17	20	
22		3	11	14
8	16	19		5

Now let's make a remark: all the magic patterns in our magic square contain one and only one empty cell: one cell is missing in each row, each column, each diagonal, each "+", "×", centre and corners, centre and the edges middle cells. If we filled in these empty cells normally, we would get 75 in each of these patterns. Now, our goal is to obtain 78, that is to say 75 + 3. If in addition to the numbers that we would put in these empty cells we add this missing 3, we increase all these patterns by 3 and then obtain this coveted 78! In other words, in the red cell, instead of putting 23, we put 23 + 3 = 26, then we continue our journey with 27; 28; 29 and 30. The trick is done:

27	7	10	13	21
15	18	*29*	4	12
6	9	17	20	*26*
22	*28*	3	11	14
8	16	19	*30*	5

It is clear that this little trick requires a little practice. It is advisable to start by getting used to simply filling the square, always starting from the same cell, and always using the same variation. Then, to get used to the magic constants that are multiples of 5, and finally to practice changing the last five cells.

The sequentiality of the numbers is unfortunately lost if the requested magic constant is not a multiple of 5. However by not announcing what is being done in advance and by revealing all the magic patterns one by one, as done earlier in this section, the effect is real and the well-trained reader will pass for a genius. Like any magic trick, you should avoid repeating it, unless of course you have practised to do it again with the other variation than the one you used the first time.

Teaching ideas

Obviously, the work on multiples, divisors, similar to that of finding our base number from the previous part, is similar to finding the starting number of this part. We can also organize a contest to see who will fill the magic square the fastest with an imposed magic constant.

More simply, we find in the year 8 textbooks of the Sésamath mathematics teachers association the following magic square:

		0	8	
		−11	2	
−9	−1	12		3
−3		−12		9
−2	11	−6	7	

The goal of the exercise is to complete the magic square, knowing that the magic constant is 0 and that all the numbers from − 12 to 12 are present. An exercise on relative numbers, but not only, since the students have an additional obstacle with the four cells at the top left, which will also require a little logic. The reader is encouraged to try to solve this little puzzle. In case of problem, we can recognise the second variant of the method, starting from − 12.

Yann Le Bail's magic squares

The last magic squares of this book are a combination of panmagic squares of order 5 and the method of Professor Roger Bowley.

Indeed, to create his square, as we have seen, Roger Bowley used four digits having symmetries, based on the base 4, in a 4 × 4 square. Unfortunately, a panmagic square of this kind does not exist. Is it possible to achieve the same feat but this time of order 5, and, if possible, panmagic? The answer is yes.

The method

All we need to do is use the following five digits: 0, 1, 2, 5, 8. By allowing ourselves to write numbers systematically with two digits, we can obtain, for example:

10	25	51	88	02
81	08	12	20	55
22	50	85	01	18
05	11	28	52	80
58	82	00	15	21

Here, we started in the middle at the bottom and we followed the first variant of construction of the panmagic squares. The second variant obviously also works, but has the disadvantage, in my opinion, of creating a column with the numbers 00; 11; 22; 55; 88, which I find less aesthetic...

Thus, from now on, by turning upside down, or not, this magic square, by looking at it, or not, in a mirror (or through the paper), we always obtain a magic square, of magic constant 176, which broken diagonals, *going around*, have a sum of 176; which centre and four corners have a sum of 176; which centre and the edge middle cells have a sum of 176; of which, if a number is chosen at random, the pattern "+" or the pattern "×" with this number at the centre has a sum of 176 (if the number chosen is on an edge, we *go around*).

In short, it seems that no matter what we do, we will always get 176! I have tears in my eyes... Doesn't this magic square represent the ultimate in a magic square? Isn't it inherently perfect? Isn't it the real answer about life, the universe and everything? Am I overdoing it, by any chance?

Teaching ideas

A question is left to the reader and proposed to the students: For each of the two methods, there are 25 possible starting points.

That is to say that in all, it is possible to create 50 different Le Bail's squares (this too can be given to the students to discover).

Take one of these Le Bail squares. By turning upside down or mirror effect, we find three other Le Bail's panmagic squares. Are these three obtained squares part of the original 50 or are they new Le Bail's panmagic squares? So, how many Le Bail panmagic squares are there?

Conclusion

Ainsi s'achève ce livre. Celui-ci est bien sûr loin d'être exhaustif sur toutes les méthodes pour créer des carrés magiques. On pourra par exemple citer la méthode LUX de John Horton Conway pour créer des carrés magiques d'ordre multiple de 2 mais pas de 4.

Les méthodes de ce livre, à part la multiplication de la partie 5, sont des méthodes directes, c'est-à-dire qu'elles ne nécessitent pas de dessiner plusieurs carrés intermédiaires. Il existe bien d'autres méthodes, directes ou indirectes, que ce soit pour les ordres pairs ou impair. Le soin est laissé au lecteur de les découvrir.

Merci de m'avoir suivi jusqu'au bout dans ce petit voyage non exhaustif au cœur des carrés magiques.

Thus ends this book. Which is of course far from being exhaustive on all the methods for creating magic squares. For example, we can cite John Horton Conway's LUX method for creating magic squares of orders multiple of 2 but not of 4.

The methods in this book, apart from the multiplication in part 5, are direct methods, that is, they do not require drawing several intermediate squares. There are many other methods, direct or indirect, whether for even or odd orders. It is left to the reader to discover them.

Thank you for following me to the end in this short, non-exhaustive journey into the heart of the magic squares.

My special thanks to Daniel Argueta for proofreading the English version of this book.

Bibliography

Association Sésamath. (2016). *Sésamath: Le Manuel de cycle 4*. Magnard.

Bégin, C., & St-Hilaire, C. (2021). *Carrés magiques – nouveaux horizons*. http://collections.banq.qc.ca/ark:/52327/bs4353355

Bowley, R. [Numberphile] (21st February 2012). *Special Magic Square* [Video]. YouTube. https://www.youtube.com/watch?v=aTSYARnB-3Y

Denef, Y. (2019). *Le problème du cavalier d'Euler*. Le problème du cavalier d'Euler. Retrieved the 3rd November 2021, from http://ydenef.free.fr/

Gardner, M. *The Colossal Book of Mathematics : Classic Puzzles, Paradoxes, and Problems* (1st ed.). (2001). W. W. Norton & Company.

Grime, J. [singingbanana]. (18th January 2010). Response : Magic Square Tutorial [Video]. YouTube. https://www.youtube.com/watch?v=tTCzcBs2b5Q

Grime, J. (2012). *Non-transitive Dice*. James Grime - Mathematician, lecturer, public speaker. Retrieved 10th May 2022, from https://singingbanana.com/dice/article.htm

Hartley, M. *Making Big Magic Squares*. Dr Mike's Math Games for Kids. Retrieved 26th July 2024, from https://www.dr-mikes-math-games-for-kids.com/making-big-magic-squares.html

Heinz, H. (2009, September). *Magic square Update-2009*. Recmath : Archive of recreational mathematics. Retrieved 31st October 2021, from http://recmath.org/Magic%20Squares/square-update.htm

de La Loubère, S. (1691). *Du Royaume de Siam : Tome 2*. Coignard.

Le Bail, Y. [TyYann]. (7th August 2020). *Magic Squares* [4 videos]. YouTube playlist : https://www.youtube.com/playlist?list=PLV_8Ld73yCWEJlCA4LQrioGYgMCn4gN7x

Sesiano, J. (2005, April). Les carrés magiques en terre d'Islam. *Dossier Pour la Science, 47*, 38-41.

Stertenbrink, G. (2019, 26 November). *Computing Magic Knight Tours*. Computing Magic Knight Tours. Retrieved the 3rd November 2021, from http://magictour.free.fr/

Van Den Essen, A. (2016). *Carrés magiques : Du Lo-Shu au sudoku, comment un casse-tête vieux de 5 000 ans a conquis le monde*. Belin.

Villemin, G. (2016, 9th December). *Carrés magiques, ordre 4 pandiagonaux*. NOMBRES - Curiosités, théorie et usages. Retrieved the 3rd novembre 2021, from http://villemin.gerard.free.fr/Wwwgvmm/CarreMag/CMor4Pan.htm

Villemin, G. (2021, 17th March). *collection de nombres, carrés magiques, ordre 3*. NOMBRES - Curiosités, théorie et usages. Retrieved the 31st octobre 2021, from http://villemin.gerard.free.fr/Wwwgvmm/CarreMag/CMordre3.htm

Wikipedia contributors. (2021, 22 September. *Carré magique (mathématiques)*. Wikipédia. https://fr.wikipedia.org/wiki/Carr%C3%A9_magique_(math%C3%A9matiques)

Yabuuti, K. (2000). *Une histoire des mathématiques chinoises (Regards)*. Belin.

www.ingramcontent.com/pod-product-compliance
Lightning Source LLC
Chambersburg PA
CBHW051534240526
45471CB00020B/2561